Dialogo

DIALOGO

Primo Levi   Tullio Regge

Translated by
Raymond Rosenthal,
with a New Introduction by
Tullio Regge

Princeton University Press
Princeton, New Jersey

Copyright © 1989 by Princeton University Press
Published by Princeton University Press, 41 William Street,
Princeton, New Jersey 08540
In the United Kingdom: Princeton University Press, Oxford

Italian edition copyright © 1984, Edizioni di Comunità, Milano

Library of Congress Cataloging-in-Publication Data

Levi, Primo.
  Dialogo.

  I. Regge, Tullio.   II. Title.
PQ4872.E8D513 1989   853′.914   89-10325
ISBN 0-691-08545-5

This book has been composed in Linotron Bodoni

Princeton University Press books are printed on acid-free paper,
and meet the guidelines for permanence and durability of the
Committee on Production Guidelines for Book Longevity of the
Council on Library Resources

Printed in the United States of America by Princeton University Press,
Princeton, New Jersey
10 9 8 7 6 5 4 3 2 1

Contents

Introduction

I never remember dates and I hate to keep notes, and unfortunately as a result I do not know exactly when I saw Primo Levi for the first time. I do remember, however, that it happened about a dozen years ago at my home in Turin, and the occasion was an informal dinner with the mathematician Corrado Bohm, who at that time was my colleague at the University of Turin, and his wife. I had invited the Bohms, and Corrado suggested bringing along Primo, and I was only too happy to concur.

Primo came without his wife because she was busy. After we introduced ourselves, he pulled out from a box a bunch of flowery contraptions made of steel wire and some gaudily colored plastic film. Right away I said: "So you bought them, too."

He was terribly disappointed. "Bought? I made them with a wonderful new invention of mine—a plastic compound. You could not possibly have seen them before."

"You can buy the kit for a couple of dollars in every variety store in Princeton. They use it to make horrible artificial flowers, but I bought it anyway," I said.

"But why?"

"Because they illustrate a beautiful mathematical problem, the problem of plateau, that of constructing the surface of minimal area spanning a given contour. The contour is here given by the steel wire and the surface by the plastic film stretched on the wire. You can get the same effect with soap films, but those you cannot touch."

This was exactly the reason he had developed his formula. He forgot his initial disappointment, and our common interest in the subject bloomed into a very exciting conversation. Primo was a first-class writer but never forgot his roots as a chemist. His scientific culture was broad and well balanced, and he possessed an extraordinary gift in explaining things clearly—in plain but nevertheless very clear language.

We continued to talk about science during dinner, and it is a pity that I do not remember in detail all that was said. Halfway through our meal, he stopped and quite casually said: "By the way, during the war I was in a concentration camp."

What a monumental understatement. I did not know what to reply, and I am afraid that I just uttered some harmless nonsense. But then I still have no idea about how I would have drawn Primo into a conversation on the Holocaust.

Many years later I was invited to visit Poland; I ended up in Cracow and eventually in Auschwitz. I went through the infamous gate with the "Arbeit macht frei" inscription on it; I saw the movie made by the Red Army; I listened to Beethoven's funeral march; I saw ghastly pictures; I read some terrifying statistics, and I thought about Primo. I still do not know what I would say, and, worst of all, I have to fight constantly against the common survival mechanism that edits out unpleasant facts and memories.

I am ashamed to say that I did not press Primo on the subject of the Holocaust as much as it was my moral duty to do. On the other hand, Primo was certainly not an egocentric writer, one of those who recycles himself endlessly under different labels. And maybe sometimes he wanted to get away from it all.

His cultural appetite was boundless. It was always on the go and ranged from literature in four or five different languages, through history, recent and old, Jewish culture, philology, and back to science. He was not the only contemporary Italian writer to be interested in science (for instance, the late Italo Calvino had similar inclinations), but Primo was a professional chemist and could talk on equal terms with any scientist.

He proceeded to tell me that he had worked as a student in the same wing of the physics building where my office was then located. He did so under the supervision of the Italian physicist Dallaporta, also known as "Potassium" for reasons I shall explain later.

I relaxed, I forgot about the Holocaust, and our conversation embarked upon a safer and asceptic course centered on one of his latest books, a collection of science-fiction stories which he had shyly published under the pen name "Malabaila" but which was immediately recognized by everybody as Levi's work, much to his embarrassment. I told him that I had enjoyed it, particularly the story about a vast air rescue operation of a human population doomed by famine. He was his usual modest self and replied that the central idea of the story was not all his own and that he had borrowed it from someone else.

I saw Primo again on many occasions after this, but our conversations were usually shorter and less informative. I once convinced him to come and see a curious computer animation wherein a visiting American mathematician attempted to visualize the fourth dimension. He seemed to like it, but later admitted that he was not quite sure that he had seen the extra dimension. Maybe that old fellow Kant scored some points in talking about categories.

The Italian radio and television broadcasting system (briefly RAI) invited me eventually to organize a series of about a dozen conversations on physics meant for the general public and lasting about a half hour each. The first broadcast was to be introductory and would cover the history of science, particularly of physics. The idea was to gather two or three people and let them talk freely on the subject; under no circumstances would I have agreed to the standard and stuffy academic lecture. I could not think of a better candidate than Primo. So I invited him, he accepted enthusiastically, and eventually came on the program with Carlo Augusto Viano, a friend of mine and a prominent philospher and historian.

The broadcast was very successful. Primo took his job very seriously and insisted on the utmost standard of clarity for all of us. Moreover, I think this was one of the few recorded occasions in which he spoke about science, though, of course, there are many such recordings concerning his life and literary works.

Meanwhile, I had become a collaborator and scientific writer for the daily *La Stampa*, a leading Italian newspaper published in Turin. Needless to say, I was again in good company with Primo, who also worked for the paper.

The power of mass media is awesome. It is impossible to appear on television and write for a newspaper without attracting the curiosity of people. Primo had already known this for a long time, but popularity never spoiled him, and he never allowed it to turn him into a prima donna or a monument. I have no idea as to whether I was equally successful. But at any rate, I tried to resist and to learn from others there, including Primo.

Popularity always seems to call for a book, and soon enough Ernesto Ferrero, a good friend of mine and at that time the chief editor at Edizioni Comunità, an Italian publishing house, came knocking at my door and stated flatly that he wanted a book from me. He was realistic enough to know and to declare quite openly that he considered me to be outstandingly lazy. His proposal was that I should allow myself to be interviewed by a gifted journalist, that the interview be taped, that it be lightly edited and published. He promised that I would not be asked to do any burdensome work. It sounded appealing enough, and in principle I agreed.

Then came the delicate matter of the choice of the interviewer. Ernesto had a list of names of good professionals, but none of them really satisfied me. At the end I stated half seriously that, after all, the best choice would be Primo Levi. Ernesto's extraordinary instinct made him jump at the idea, and the interview was quickly upgraded to a dialogue. He immediately contacted Primo, who agreed to the arrangement instantly.

I was hooked. I had no choice but to go forward and do my best. I had another dinner with Primo and Ernesto at the home of the publisher Franco Debenedetti, followed by about two or three recording sessions, all held (if I remember correctly) at my home. The tape became a preliminary manuscript to which Primo and I added a few comments and footnotes, but, on the whole, the ensuing book represents quite faithfully our informal conversations.

I remember these free-wheeling exchanges of ideas as a very happy time in my life, but they did not come without some feelings of guilt. Primo had truly an omnivorous curiosity about everything, particularly about science, and he put no restraint on my intellectual wanderings. He did not try to divert me from my own favorite subjects to that of his own battle of survival in the camp. As a result, I did most of the talking even when human decency would have called for prolonged silence and listening on my part.

At one time, however, he put his hand forward. The sleeve of his shirt slid back, and I saw a number tattooed on his wrist. The Holocaust was back, and again I said something silly: "I take it to be the original."

"What else? Nobody goes around buying fake Nazi tattoos."

"How many people in Turin are left with such a tattoo?"

"Not many. You can count them on your fingertips. I know well only an old lady who has it."

Conversing with him in those days aroused my curiosity about his books. I started reading *Survival in Auschwitz*, which brought him everlasting fame. I could not finish it. The Holocaust haunted me and I had to make a conscious effort to go on reading. I was more successful with *The Periodic Table* and his more recent *Other People's Trades*, both of them collections of shorter essays and stories. In the first, he associates a character to a specific chemical element. For instance, Delmastro, a very strong-willed and heroic schoolmate of his (he was shot by the Fascists), is linked to Iron. In the story he describes Iron as coming from a family of farmers, and this, Primo said, has offended the surviving relatives. Of course, Primo had meant no offense, but still the reaction of the family had pained him. He asked me: "Do you really think that I put them down in calling them farmers?" I replied: "For heaven's sake, no. I come from a family of poor farmers and from the same area as the Delmastros, and I am proud of it."

He still looked unhappy and told me that this was not the only time he got an adverse and unexpected reaction from people he had portrayed in his writings. He told me that after a few unpleasant experiences he took up the habit of showing his manuscript prior to publication to the people who were portrayed therein. Even this was not enough. A lady friend of Primo's, for example, seemingly approved of what he had said, but it was clear from her face that she did not like it.

Later on, in fact, I was warned to use, when referring to farmers in my newspaper articles, the upper-crust *agricoltore* instead of *contadino*, as Primo had done. The word "contadino" meant originally a servant, practically a slave, in the old feudal system, and even today farmers tend to resent it.

My conversation with Primo was not the end of the story. Strangely enough, a former high-school friend of mine, also named Delmastro and a man of extraordinary physical strength, indeed ironlike, came to visit me as I was revising the manuscript for this book. Acting on impulse, I asked him whether he had read *The Periodic Table* and whether he had any idea as to who that Iron-Delmastro person really was. I hit the jackpot: Iron was his uncle, and my friend was the last person in his family to see him alive before the Germans captured and executed him.

I told Delmastro about my conversation with Primo and asked his opinion about the whole story. The reply was that, really, the Delmastro family had not been annoyed at being

labeled farmers and that Primo had somehow misunderstood
their reaction. The Delmastros had been builders for generations; in fact, their name is related to their profession. They
just felt that the story was not accurate in some of its details,
including the political affiliation and ideology of Iron. My
conclusion was that Primo was above all an artist, but in no
way could he be considered an accurate historian.

I had the unique opportunity to get a direct check on another character in the book, this one "Potassium." I finished
reading the book by the time Primo and I had our last conversation, and I told him that I had instantly recognized the
character he called Potassium. It had to be Dallaporta, the
well-known and now retired Italian physicist who had held a
teaching position in Turin during the war and who had taken
personal risks in helping Primo under the racial laws when
he was still a student.

I was right. And I found out that he had fond memories
and high esteem for Potassium, but dissented from him when
Potassium, a very religious and sincere ecumenical Catholic,
had attributed Primo's survival in the concentration camp to
a direct intervention of Divine Providence. Primo very em-
phatically refused to be among the chosen ones and told me
that God had allowed the Nazis to murder people who were
far worthier than he.

At any rate, years later I met Potassium on a tourist boat
crowded with physicists who were going from Rapallo to Por-
tofino on the Italian Riviera. It was a beautiful, sunny day.
People were talking excitedly and singing. The noise was
deafening. I shouted to him: "Are you Potassium?" He
smiled and fired back instantly: "Yes. No further comment
needed."

Primo wanted to know my reaction to his other book, *Other People's Trades*. Of course I had read it, and he challenged me to tell him which of the short stories in the book had been completely fabricated. I tried to guess but failed miserably, and probably I shall never know the answer. He admitted that he had also faked other stories a bit, particularly the one in *The Periodic Table* about a German Nazi chemist who had been his boss in the concentration camp.

I dislike coming to the end of my recollections of Primo. I had wanted him to come on RAI again, this time in a program dedicated to symmetry. But his health had deteriorated, and he was forced to go through some debilitating though not critical surgery. This event left him depressed, and it became more and more difficult to convince him to come out of his home, where he also attended his old mother. During this time I kept receiving phone calls from the most unlikely people who sometimes wanted to invite him to the silliest places and expected me to act as a go-between.

One Saturday afternoon I came home from a long trip by car, due to which I had been unable to listen to the news. When I saw the agitated face of my daughter Anna, I knew that something terrible had happened. She asked, "Daddy, do you know that Primo Levi died?" Of course I did not know, and I was greatly saddened to learn it was a suicide. Or was it? Not everybody agrees, of course.

Quite recently I appeared at a meeting in Turin where Rita Levi Montalcini's book, *Elogio dell'imperfezione*, was presented to the public and discussed. Rita is a towering figure in Italian and American science, a Nobel laureate for her research on the nerve growth factor (NGF). What's more, she is also Piedmontese. The moment we saw each other, we be-

gan to talk about Primo, whom she knew well. She refused
adamantly to believe that he knew what he was doing when
he took his life and thinks that it was due to a raptus, a
temporary and sudden lack of control, perhaps an aftermath
of his medical treatments. Nobody will ever know the answer.

Upon reading her book, I had the uncanny and somehow
frustrating feeling of going through the political and cultural
life of Turin as it was before and during the war. Both Primo
and I were born and raised here, and he has devoted many
of his writings to the place.

Turin is a strange place to live in, and the least known of
the big Italian cities. As a rule, it is avoided by tourists, who
wrongly think there is nothing to see and suffer the local di-
alect as an ear-splitting parody of proper Italian. Turin is,
however, not totally deprived of redeeming features. It is
home of the sprawling Fiat industrial complex and, a little
more than a century ago, it was the first capital of Italy.
Champollion spent years here to study the wonderful master-
pieces of the local museum of Egyptian art, second in importance only to that in Cairo.

Turin lies in the midst of hilly country rich in splendid
wines, truffles, and good food. Here was the first perform-
ance of "La Bohème." Here we had Toscanini as conductor
for many years at the Teatro Regio. Here was born Louis
Lagrange, the father of modern analytical mechanics. And
here the reactionary French baron and mathematical genius
Augustin Cauchy discovered, while a fledgling from Paris,
the famous theorem that bears his name. Nietzsche loved Turin as well and lived here many years, until he went mad and
kissed a horse in the street (or maybe he was mad even be-

fore then). Jean-Jacques Rousseau came here as a young musician but hated the place vehemently. Among other visitors were Dostoyevsky and Henry James.

Italians from other regions of Italy are known to dislike Turin. They do not think of the Piedmontese as proper Italians, but rather as some sort of dull and dumb Gallic-Prussian invaders, their favorite target of Polish jokes. They have not forgotten the colonial policy at the end of the nineteenth century when Garibaldi and the regular Piedmontese army conquered Naples and took north the treasury of the Bourbons to nurture the budding industry. Yet they flock to Turin to buy cars and advanced technology and, above all, to seek advice from the legions of local magicians and crooks who make good on Turin's sinister reputation as home of the supernatural—one of the corners of the famous Lyon-Prague-Turin magic triangle. (I would not bet a penny on it, of course, but local experts claim that it is much more effective than the Bermuda Triangle.) For awhile, this reputation was strengthened by the presence in town of the world-famous Shroud of Turin, which was believed to have wrapped the body of Jesus Christ after it was taken down from the Cross. Thanks to carbon dating, the shroud has been conclusively proven to be a fourteenth-century fake.

Most of the population of Turin is now heavily mixed with immigrants who came from poorer regions of Italy, looking for good jobs. In addition, Third World countries are now contributing, sometimes illegally, to the change in ethnic balance of the place.

Primo, Rita, and I were born here (by a tragic quirk of destiny both Primo and Cavour died in the same house where they were born). All of us shared memories of the past. We saw the Second World War coming and the troops parading in the streets, and we saw the same King Victor Emanuel who was described by Hemingway in *A Farewell to Arms* and who was a Piedmontese but could not care less for us and for the rest of Italy.

Finally, we all went through the shattering experience of food rationing, of aerial bombings, and of war horrors. I still vividly remember a morning after a terrifying bombing as we came out of the cellar, saw the old part of town in ruins, and decided then and there to leave the place on bicycle. And yet I have the feeling that I was still too young and immature to really understand what was happening when it was happening. Primo was a dozen years older than I and had already reached maturity when the war broke out.

Rita ends her book with a few emotional and tense pages dedicated to Primo. At the end of her book she quotes him as saying, "I am not a prophet." I am sure that he said this in despair at the many pointless requests for public appearances and advice both in Italy and abroad. People like him are endlessly and ruthlessly cajoled into attending boring round tables and meetings, giving interviews and speeches on remote and uninteresting topics. So I can well understand why at the end he had to say, "I am not a prophet."

If by that he meant that he abhorred the arrogance and missionary zeal of prophets and adulation of sycophants, then he was right. If, however, he meant that he left no moral lesson, then I must agree with Rita that, on the contrary, he is and will always be very much with us and our children.

Tullio Regge

Dialogo

Dialogo
Between Primo Levi and Tullio Regge

Regge: Among the things I have in common with Primo Levi, and that Levi does not yet know, is a real secret mania: I'm studying ancient Hebrew, on my own. I have acquired a Bible, with a facing text in Hebrew, and now I'm almost at a point where I can dispense with the Italian text. I've already gone through all of *Genesis*.

Levi: Is this passion born from a linguistic, philological interest, or from something else?

Regge: From a philological interest. But I've found that reading the Bible is interesting in itself. Like many intellectuals of my stamp, I am considered an agnostic. It's not that from a religious standpoint the Bible sweeps me off my feet; nevertheless, it remains a fascinating document.

Levi: Not to mention the fact that if you manage to read it in the original, you can notice all the tampering it has been subjected to: the translations are in fact reworked. But I must confess that my Hebrew is certainly much poorer than yours. It's the Bar Mitzvah Hebrew of the religious majority that one learns at thirteen and forgets before eighteen! Too bad.

Regge: Among those tamperings I have noticed the episode of Joshua's two spies who go to Jericho and are taken in by a woman whom the text calls *zoná*, that is, prostitute. Now, this does not appear in the Italian translation; they have preferred to use euphemisms.

Levi: That's Rahab, whom Dante later put in the *Paradiso*, among the loving spirits, precisely because she favored the Hebrews' entry into the Promised Land, and because she was the progenitrix of David and therefore of Christ. She is also mentioned in the Talmud: in the Talmud there's everything, as is well known. There is also the story about seductresses, bewitchers. There are five of them, if I'm not mistaken. One, I think, was the "manly Jael" who seduced all those who heard her voice. Another seduced by contact, with her clothes or with her hair. But the most extraordinary was in fact Rahab: every man who pronounced her name immediately ejaculated semen.

Regge: The Talmud is a boundless work; I cannot say I have read it all. I believe that this interest was transmitted to me by my father, who had bought many prayer books from the book stalls, many years ago. The first book contained abstracts of Rabbinical judgments, which I read to get the feel of it. They are antifeminist in the most incredible way.

Levi: In this regard, too, there is no doubt that the Talmud contains everything and the contrary of everything. It depends on what filter you use. You can extract from it feminist judgments as well as antifeminist judgments, praise of study as well as its abomination. You can indeed find everything in it. Also nonsense, such as the story that the Eternal Father spends three out of twenty-four hours studying the Torah, that is, himself.

Regge: The complexity of the work and the cross references are such that studying it seriously is an undertaking that demands a lifetime: you have to be a real professional to do it. As for myself, I realized afterwards, thinking about it, that the study of ancient Hebrew serves as a polemical reply to my classical studies: almost the need for a negation of Latin, which accompanies the continued necessity to broaden one's horizon.

Levi: My Hebrew progenitors had only half understood the alphabet. Compared to the Greek alphabet, the Hebrew use of the alphabet is quite irrational, to say the least, and it still is, even in modern Hebrew. There are three mutes, but anyone who writes an "aleph" instead of "he" or an "ain" is considered an ignoramus. The vowels must be guessed at. . . . The impression we get today is probably due to the fact that the pronunciation has changed since those days. In the past these mutes were not mute. Why indicate a mute if it's not there? It's a bit like what happened to the "h" in French: at one time it was pronounced, it had a function. In Polish there is an intermediate consonant between the "s" and the "sh," an "s" with an accent. The Polish have very

many sounds, all of them highly specific. It is difficult for us to distinguish among them. All peoples
specialize.

Regge: I bought a human voice synthesizer for my micro-
computer. It clearly enunciates numbers, and also
letters, some quite felicitously, others not so well. It
is also possible to compose words, which is of
course its purpose. But the "r's" are difficult to syn-
thesize. Getting the computer to say "Maria" is
quite an undertaking. On the other hand, you cannot ask for more from a device the size of a music
cassette. What's difficult is to get it to keep quiet.
. . .

Levi: At Turin Radio they took one of my stories, "The
Versifier," and they got a synthesizer to tell it. They
made me a present of the disc, in which there is this
little machine that introduces itself, says its name,
tells its own story: "Once upon a time I spoke in a
monotonous manner, on a simple note, but now as
you can hear I'm able to do much more." It recog-
nizes punctuation marks, and distinctly lowers its
tone before a period, lowers it a bit less before a
semicolon. It does have a certain ability to articu-
late, but the final effect is metallic, unreal.

Regge: To get back to Hebrew, that isn't the only interest
my father transmitted to me: I also owe him my scientific vocation. He was born a peasant, very poor,
in the plain at Vercelli, near Borgo d'Ale. But he
was extremely clever; he must have had an excep-
tional I.Q. By working like an animal, he managed

to save sixty lire, which he used to obtain a surveyor's diploma. This is during the First World War, and in order to take his final exam he flung himself into a desperate race on trains that shuttled back and forth from the front lines. He managed to get his diploma, which in those days was very important. He turned out to be the only surveyor in an area that embraced a population of several thousands. This marked a notable improvement in his economic situation. He came to Turin, got married. He had the autodidact's broad-ranging interests, obviously chaotic. At a certain point he set out to write a book on astronomy in which he maintained that the force of gravity was the inverse of the cube and not the square, and therefore Newton was wrong.

Levi: But how did he demonstrate this?

Regge: By making a mistake, but the mistake was quite interesting. He had discovered that the height of the tides induced by moon and sun are approximately the same. If you do your calculations, this means that the height of the tide behaves as the inverse of the cube of the distance, which is true. The tides behave essentially as the derivative of the force of gravity. What produces the tide is not so much the force of gravity in itself, as the way in which it varies. The trouble is that he made the mistake of thinking that the tides were in fact proportional to the force of gravity and not the derivative. But I was never able to explain his mistake to him. . . . My

father had learned everything by himself, trigono-
metric functions, algebraic equations, a certain
mathematical formalism. He browsed through the
bookstores and bought university course papers by
the kilo, and so I came to own these texts that go
back to Boggio's time, still written in longhand, lith-
ographed. And so I learned differential calculus
when I was about fifteen. When I got my baccalau-
reate I already knew Fourier's series, integrals. But
my scientific interest began with a strange Hoepli
manual: *Delectable and Curious Mathematics.*

Levi: That was a manual to be found in every home. Its
author was Ghersi, a very strange man, who died not
too many years ago, extremely old, after giving birth
to an endless series of textbooks, and in particular
a do-it-yourself book, *New Industrial Recipes*, that I
still have, and in which he gave suggestions for the
use of certain materials—glue, glass, lead—analyt-
ically, but also explained how one can prepare oneself spiritually and physically so as to be in a state
conducive to making an invention.

Regge: The Ghersi accompanied me from eight to twelve. It
contained all these algebraic curves. I didn't understand too well what they were, but I found them very
interesting. At the beginning there were the magic
squares, the various problems about goats and cab-
bages. Despite its disorder and the book's lack of
depth, I remember it with great gratitude.

When I got to the third grade in grammar school,
my father, in an act of pure madness, made me skip

to the first year of high school. That year they had created the general high school, and this caused certain disparities of which one could take advantage. I went to general high school for two years, learning the little Latin I know, then came the bombings, and I ended up in a priests' seminary, in the vicinity of Cigliano. My interest in mathematics had by then become exclusive. Then I attended *liceo*, devoting to literary subjects the indispensable minimum of study, and perhaps not even that, since I kept having to take remedial exams. I was passionately interested in chemistry. I had found a book that told the story of the elements. . . . Translated from German, a classic, at the university level. You have certainly seen it. I was interested in the periodic table; that's why when your book came out I pounced on it. I liked the sentence in which you say that the periodic table is poetry, and besides it even rhymes.

Levi: The expression is paradoxical, but the rhymes are actually there. In the periodic table's most common graphic form, every line ends in the same way with a hallogen plus a rare gas: fluorine + neon, chlorine + argon, and so on. But in the sentence you are quoting there is obviously more. There is the echo of great discovery, the discovery that takes your breath away, the echo of the emotion (also esthetic, also poetic) that Mendeleev must have experienced when he intuited that by ordering the elements then known in that particular way, chaos gave way to order, the indistinct to the comprehensible: it became possible (and Mendeleev did this) to identify empty slots that had to be filled, since "all that can exist exists"; that is, to do the work of prophecy, foresee the existence of unknown elements, which later on were all punctually discovered. To discern or create a symmetry, "put something in its proper place," is a mental adventure common to the poet and the scientist.

Regge: During the war, in order to survive my father found a job at the Technical Office of Venaria township and we all moved there. We actually lived in a castle with very thick walls, a park in front, and I had the castle all to myself. On the 8th of September (1943) the mob broke in, looting everything. Well, not precisely: for example, they hadn't noticed the rooms of the army pharmacy, which were an inexhaustible store of strange chemical compounds, such as sulphide of antimony, biodure of mercury, which was a beautiful bright red that, when heated, turned yellow. . . . Iodine in vials, carbon tetrachloride with chlorine dissolved in it, compounds for tear gases, kilos of methylene blue, picric acid.

Levi: That's an explosive for detonators.

Regge: Yes, I played with dangerous stuff, potassium chloride, smoke grenades. I set fire to everything. One very beautiful reaction was to mix arsenic anhydride, which is somewhat poisonous . . .

Levi: It's very poisonous: it's Madame Bovary's poison.

Regge: . . . with potassium chloride and sulphuric acid, getting a highly oxidizing mixture, heptoxide of chloride which at a certain point of concentration becomes green, and then it explodes. And in fact the tube popped and shot the mixture onto the ceiling, where it began to corrode the plaster. Then I mixed bromide with sulphuric acid, I heated it, I created bromine; the bromine began to evaporate, chewed up all the rubber tubes that connected the glassware, inside a minute. I filled the house with bromide, my mother arrived. . . . I was a domestic danger. Otherwise, I roamed through the cellars of the castle, which I remember very well, carrying one of those resinous torches that never go out. From cellar to cellar I descended three stories under the earth, until in one room there literally exploded ten thousand bats, which slept there after having entered through the cellar's shaft. I managed to escape; I never went down there again. I returned to my experiments and survived them. Once I swallowed some lithium . . .

Levi: Swallowed deliberately?

Regge: My father said that lithium is like sodium. Ordinarily I used it to color the torch's flame a beautiful crimsom red, but I tasted it once. I think they use it for depressive states.

Levi: Your chemical vocation somewhat precedes mine. I began around the age of fourteen. My father had certain traits like yours, but he was an engineer, and from a well-to-do family. My paternal grandfather was a small landowner; it seems he had a bank, which later failed. My father also exerted cautious pressure to orient me in a scientific direction; he too was a bibliophile, bought books at random, and had the passions of an autodidact. On his own he had studied many things, and he continued to study until the end. He had filled the house with strange books, which in part I still have. He had specialized in engineering at Liège and had found work in Budapest, where he learned German. When the First World War broke out they expelled him in a very civilized manner, even paid for his trip home. Given the epoch, it was quite rare for an Italian bourgeois to know distant countries well and speak foreign languages. For me he bought the beautiful Mondadori series of popular science, *The Microbe Hunters*, *The Architecture of Things*, a first book on genetics that was still in the process of being born—we're at the beginning of the thirties—, Carrell's *Man the Unknown*, which was published by Bompiani, and *An Introduction to the History of Human Stupidity* by Wilkins, I think.

My father hated nature. He had a savage hatred for the countryside, which to him meant staying locked up in the house without ever sticking out his nose, because there were ants, dust, because it was hot. The few times we managed to get him to take a walk, he brought along two books stuffed into his pockets and as soon as he arrived at the destination, instead of looking at the panorama, the mountains or the sea, he sat on the ground, on a newspaper so as not to dirty his suit, and pulled out his books.

Regge: My father also felt the hills were enemies. His home was a house in the center of the city.

Levi: This can be explained by the fact that since both were born in the provinces, they strongly felt the fascination of the city. My father was in love with the center of Turin. He took me there, even if I was reluctant, and he could not understand the fact that I went into the mountains to ski. Tennis yes, because it wasn't dangerous and was played within a circumscribed area. But to him the mountains were incomprehensible. He advised me to drink, smoke, go with girls. Now, I didn't smoke, I didn't drink, I had no girls. There wasn't much understanding with my father. I was substantially a romantic, and also in chemistry it was the romantic aspect that interested me. I hoped to go very far, to the point of possessing the universe, to understanding the why of things. Now I know it doesn't exist, the why of things, at least that's what I believe, but then I really believed in it. And yet I wasn't religious; religion said nothing to me, and at bottom also classical culture did not give me much. I suffered it with a certain intolerance, even though I was a good student. I had a curious sensation: that there was a plot at my expense, that family and school kept something hidden from me, which I went looking for in the places that were reserved for me: for example, chemistry or also astronomy. My father, who in this was very liberal, had bought a cellar full of books from an old man in Bardonecchia, in which there was everything, from Voltaire to Camille Flammarion.

Regge: Flammarion's books are very well written and they are still relevant, especially *L'astronomie populaire, Les étoiles*. . . . I too thought there was a plot, and that was what was taught in school. I reached the scientific *liceo* believing it would be what it was called: in fact, I built a myth round it. Instead I discovered that there they taught everything *but* science. Indeed, science was taught so badly as to discourage even the best disposed. It wasn't the fault of the teachers, of whom, everything considered, I still have a fond memory. For me it was a provocation that in the last year the hours devoted to scientific subjects were even less, to the advantage of humanistic subjects: for example, from three to four hours of philosophy. Besides, science ended with the nineteenth century, essentially. About atomic physics not a word, and it was already the end of the forties.

Levi: We were right. In my time the conspiracy was acclaimed. It was the Gentile conspiracy. I too had an excellent relationship with my Italian teacher, but when she publicly said that literary subjects have a formative value and scientific subjects have only an informative value, my hair stood on end. This confirmed in me the idea that the conspiracy existed. You young Fascist, you young Crocian, you young men grown up in this Italy must not approach the sources of scientific knowledge because they are dangerous. I never read Croce on this subject, but apparently he considered them pseudosciences, technical matters, useful for life but not for the understanding of the world.

Regge: He called them pseudoconcepts, fictitious concepts of practical utility.

Levi: This irritated me. I had two or three fourteen- or fifteen-year-old friends and we reciprocally preached these things to each other: we have found the right path, we have found the shortcut, which the school denies us. Even though we digested Greek and Latin diligently, even gladly, since we had fun linguistically, for philological reasons, as you said earlier.

Regge: I would have studied Latin if they had presented it to me in its philological aspect, but this too was completely denied me.

Levi: If they had given us Lucretius, to translate . . .

Regge: But Lucretius is difficult . . .

Levi: Or Vitruvius, or Celsus. . . . Celsus is very interesting, he explains how tonsils were operated on in his day. When you read him you realize, as I realized ten years ago, that Latin was a spoken language, not a language only good for orators.

Regge: In *liceo* the ideal Latin was Cicero's Latin, crystallized during a certain epoch.

Levi: Good for commemorative plaques, good for marble.

Regge: Latin created allergic reactions in me which by now are irreversible. It should be easier for me than Hebrew, but if I pick up Virgil I understand him much less well than a passage in *Genesis*. The only encounter that gave me a great deal of information took place during my first year in university when I read Gauss's *Disquisitiones.* Those on arithmetic and those on curved surfaces, two fundamental texts. Gauss interested me very much, and so I had to struggle through it, but I was helped by the formulas. School gave me nothing. With the old telescope my father bought me in some flea market or other, I learned to recognize all the constellations; I observed double stars, anular nebulae, the planets, the galaxies. My teacher of natural sciences was not in the least interested in all this.

Levi: As for *my* teacher of natural sciences, chemistry was a textbook, and that's it. It was pages in a book. She had never in her life touched a crystal or a solution. It was knowledge transmitted from teacher to teacher without ever a practical test. There were experiments in class, but they were always the same. They absolutely lacked everything that is inventive in such things. My teachers were decent people too, but they were totally devoid of scientific training, even though they had some sort of education.

Regge: Very true. This putrefaction of teaching was evident in the courses of natural sciences. I had made an herbarium, but I realized that my initiative left my professor absolutely cold, to the point that I soon learned to leave my fine little notebook at home. I occupied myself with it on my own, without outside input. Instead, I continued to hear that I was an ignoramus because I didn't know who Vincenzo Monti was. Today perhaps I could return with interest to certain authors, except for the fact that they give rise to persecutory memories that block me.

Levi: Did you ever try to reread *Orlando Furioso?*

Regge: Yes, but I read it without the literary smugness with which they tried to dish it out to me. I read it as a comic book. Perhaps influenced by Dore's illustrations.

Levi: It's that, too. Ariosto was a very witty man; he knew very well that he was writing a comic book romance.

Regge: Passing the maturity exam was the moment of supreme happiness in my life. The idea of no longer having certain obligations made me jump with joy. I walked on a pink cloud. Absolute liberation. I enrolled in the Polytechnic, then after the two-year course I transferred to physics because there was too much drawing. Get a serious degree, my father kept saying. Physics isn't serious. If you want to do physics, get a degree in chemistry too, because put together they are like a degree in engineering. And when I got my degree in physics with the highest marks and I was given a teaching fellowship, he still insisted. At a certain point I went to Russia and *Pravda* published my photograph. I cut it out and sent it to my father. "So he'll stop telling me to get a degree in chemistry," I explained to my Russian friends who asked me why. This anecdote is still in

circulation even now. I always run into somebody
who asks me if my father is still insisting.

The fact is that I had a mad enthusiasm for relativity and quantum mechanics, which I was beginning to discover. I had true moments of trance while
studying relativity, or certain pages of Persico's old
text on quantum mechanics, which wasn't really all
that sublime. Those were splendid years. I learned
many things—it was the dilation of knowledge, the
explosion after the *liceo's* compression. It is through
chemistry that I got to Fermi. Fermi received the
Nobel Prize for having discovered the transuranium
element, even if this is not quite exact: he developed the physics of the neutron. In any case, toward
the end of the thirties there circulated a strange review called "The Journal of Marvels," a review of
scientific culture, very naive. It is there that I read
an item: in Rome Enrico Fermi had discovered element 93. The periodic table stopped at 92; that
someone should have discovered 93 was for me a
sensational event. I mulled over it for a long time:
what might it be, what properties might it have. . . .
And I didn't know that from that moment on a new
series of rare earths was starting. Then it happened
that one day, while I was at Borgo d'Ale, in the pi-
azza, talking with the cooper, a very cultivated person, somebody came running with the newspaper,
very excited: "Do you know," he said, "the Americans have exploded an atomic bomb." Among those
who had collaborated in its construction there was
also Enrico Fermi. I saw the world of physicists as
a mythical world, which belonged to another planet.

Levi: It's clear that you were in search of answers and of
          why's. Not the last why but those just before the
          last. You were moved by a curiosity that was the
          same as mine. My father went to temple on Yom
          Kippur because he was a bit superstitious, but he
          was friendly with Lombroso, Turin's positivist phys-
          iologist; he attended mediumistic sessions, not be-
          cause he believed in spirits but so as to understand
          what was at the bottom of them. He hadn't bought
          me a telescope, but a microscope with a magnifica-
          tion of 250, which I used to organize a "classical"
          performance for him, a solution of alum in which
          you could see the crystals. . . . I had a small pro-
          jector made by Pathé Baby, very low gauge: I would
          invite my friends and put in a slide instead of film;
          you could see the crystals grow. For me, too, the
          university experience was liberating. I still remem-
          ber Professor Ponzio's first chemistry lesson, from
          which I got clear, precise, verifiable information,
          without useless words, expressed in a language that
          I liked enormously, also from a literary point of
          view: a definite language, essential. And then the
          laboratory: every year he included a laboratory ses-
          sion. We spent five hours there; it was a big com-
          mitment. An extraordinary experience. In the first
          place because you worked with your hands, liter-
          ally, and it was the first time this had happened to
          me, never mind if you scalded your hands or cut
          them. It was a return to the origins. The hand is a
          noble organ, but school, all taken up with the brain,
          had neglected it. And besides, the laboratory was
          collegial, a center of socialization where one really

made friends. As a matter of fact, I remained friends with all my laboratory colleagues. It was the team work, which in the preceding school was unknown (but I didn't know whether outside the chemistry course it is equally intense). Making mistakes together is a fundamental experience. One participated fully in the mutual victories and defeats. Qualitative analysis, for example, in which they gave you a bit of powder and you were supposed to tell what was in it: not to realize there was bismuth or to find chrome that wasn't there were adventures. We gave each other advice, we sympathized with each other. It was also a school of patience, of objectivity, of ingenuity, because the methods they suggested to you to perform an analysis could be improved: it was up to you to take a step forward on your own, to simplify. As for the last year, organic preparations, that was high acrobatics: we had to prepare a certain substance on the basis of a formula; or they might give you a crystalline substance and on the basis of tests you were supposed to say what it was, what its structure was.

For me there was a complication, as I have already recounted in my book: the racial laws. The liberation of the university coincided with the trauma of being told: watch out, you are not like the others, indeed you are less worthy then they; you're avaricious, you're a stranger, you're dirty, you're dangerous, you're perfidious. I unconsciously reacted by accentuating my commitment to my studies. I received my degree *cum laude* and I'm convinced that this praise was given me 40 pecent through my own merits and for the rest because the professors, almost all of them vaguely anti-Fascist, had found this a way to express their dissent. From my observatory it was very easy to understand whether a professor was a baron, as one says nowadays, or a scientist. But with the exception of one case, all of them were decent men. A lot of them went in for crypto-anti-Fascism, for example Petrucca at the Polytechnic. I wanted to present an experimental thesis. It seemed to me that I was entitled, since I had a very good average. I knocked on several doors. Ponzio, a notorious anti-Fascist, told me that he would have been glad to take me on, but he couldn't, the law forbade it. In the end I landed with Dalla Porta, who said to me, "Very well, laws or no laws, what does it matter?" And so I did my thesis in physics: that is, a twenty-page compilatory thesis in chemistry and an experimental subthesis in physics one hundred pages long. The real thesis was the subthesis. I should also add that among all my fellow students, men and women, there wasn't one who called me "Jew." They all saw the racial laws as a stupidity or a cruelty, or both. And yet, naturally, all of them were members of the GUF (Gruppo Universitario Fascista), the Fascist student organization. Nor was there anyone who gave signs of caution in associating with me.

Regge: Mine was a more solitary experience. Among the people to whose influence and charisma I responded

most strongly there was Professor Wataghin. Wataghin's life is truly adventurous. He was born in
Odessa, his father was a railroad employee, his
mother was a Tukashevsky. When the Revolution
broke out, the Wataghins were in Kiev, and since
two of their four sons were officers, the entire family
was in serious danger. They caught a train, arrived
at the Black Sea, and found a British ship that took
them to Greece, where they almost died of hunger.
They pushed on all the way to Belgrade, where the
father died. The sons scattered. One of them, who
in Russia had been a prominent lawyer, came to
Rome in search of work; but he had a very hard
time, until one day in the street he was recognized
by Tatiana Pavlova, who in Russia had been a client
of his. Pavlova took him to Turin, which in the twen-
ties was still the cinematic capital, as the manager
of a movie company. This brother called the rest of
the family to Turin, but when they all got there the
company went broke. Only Gleb, the eighteen-year-
old, remained in Turin and was greatly helped by
Corrado Segré. Around 1925 it was young Wataghin, who in the meantime had been offered Italian citizenship, who brought to Turin the classical
texts of atomic mechanics and relativity, until then
unknown. The Italian government then sent him to
Brazil for a series of cultural exchanges, and he returned to Brazil several times, meeting on different
occasions with such personalities as Rachmaninoff,
Aliokin, Ungaretti. In Brazil he founded a school of
physics which to this day bears his name, and he

returned to Italy at the end of the war. More than a calculator, he was a man of great intuition: his very presence magnetized the institute; he attracted important guests—Virac, Pauli. Another rare quality of Wataghin's was knowing how to detect at first sight whether a person was mediocre or not.

Levi: Perhaps that's not so difficult, given certain parameters. Certainly one must know how to set aside feelings of sympathy or antipathy. Above all, sympathy. Sympathy is often a means of getting away with incompetence. Italy is full of attractive people who are very *simpatico* and don't know their trade. Being able to evaluate the person in front of you is a basic operation—to weigh him—I imagine that also among physicists there are those who play the part.

Regge: It's really very easy. I sometimes do it for a joke; I amuse myself by posing as an expert in areas about which I know absolutely nothing. I find it incredibly easy. All you have to do is find out what the key words are. As an experimental physicist I'm not worth much. One time I got somebody to explain to me what one is supposed to say to a person who builds accelerators. There are a couple of phrases, tune-shift, increasing the injection current, field stability. . . . I learned them and one time at a dinner I trotted them out, I went on for twenty minutes.

Levi: Everybody knows that having insight into people is the key to any career. They say that is how Napoleon made his career.

Regge: Also Cottolengo, I think, a personality whom Togli-
atti admired very much. I once read an obviously
partisan biography in which it was explained that he
had known how to choose for himself a circle of very
able collaborators, and this ability in choosing them
must be counted among his less visible but more
certain merits. He might have been as charitable as
you wish, but if he hadn't been a good administra-
tor. . . . Well, after the university I went to a summer school at Les Houches, near Chamonix, which
to some extent was a prototype of its kind. More
than by the courses, I was changed by the personal
conversations. I learned English and French, I
talked with physicists of a very high international
level. In '54 I went to America for the first time at
the suggestion of Caianello's, and inside my head
there were images of the Oppenheimer trial, which
had exploded that summer. Oppenheimer was part
of my myths, like Fermi. When I arrived in Rochester they immediately put me through the wringer.
The Americans are serious people, perhaps a bit
boring, but rightly exigent on the professional level.
That contact was decisive. Not that I uncondition-
ally adore America. Certain American things I don't
like at all and I don't want them. I learned what it
means to live abroad, and to try to understand the
mentality of one's neighbor. At any rate, in Rochester I met my future wife, she too a physicist, she
too having landed there through an application anal-
ogous to mine.

   As for the Oppenheimer case, I fully shared the

attitude of the physicists of that time: science is by
its nature neutral, and so the scientist bears no responsibility. Now I think in quite a different manner, more nuanced. At that time the physicists
closed ranks. Two types of problems converged in
the Oppenheimer case: the responsibility of the scientist who builds lethal weapons, and the scientists'
loyalty to the American nation. On the second problem the physicists were divided by a great rift.
There were those who maintained that Oppy had
passed information to the Russians, and perhaps
even continued to do so, something that I don't believe. I hadn't been able to see Oppenheimer then,
just as I didn't meet Fermi, who died a few months
after my arrival in America, or Einstein. But Fermi
was at the height of his activity, Einstein was already very old and outside of modern physics: at
least two generations of scientists had already
passed. I think I did well not to see him. I would
have seen only a monument; I would have violated
his dignity by going to look at him precisely as at a
monument.

In Rochester I managed not to do laboratory work
in experimental physics, for which I had a real aver-
sion. Contrary to what was the case in Turin, in
Rochester the laboratory was very sophisticated, full
of ultramodern equipment, a century ahead in comparison to ours. I had gotten together with a student
who knew nothing about theoretical physics, but in
recompense he knew how to handle the various in-
struments. Our collaboration went off very well and

would have continued, but the department decided
to separate us. At the end of my two years I was
given my Ph.D. anyway, even though I hadn't done
any laboratory work, and I returned to Italy, passing
through Leyden, where Wheeler initiated me into
the mysteries of the black hole. Wheeler is another
incredible personage. He had (still has) a mania for
explosions, from firecrackers that he bought by the
kilo, right down to the atom bomb that he had seen
explode by working on the Manhattan Project.
While we were speaking about black holes, he suggested that we study the oscillation of the gravitational field around the black hole. He was a theoretical physicist of the first order, and he influenced
me greatly with regard to relativity. When I returned
to Italy in 1956, we continued our collaboration by
mail, then he invited me to Princeton. But meanwhile I had spent six months at the Max Planck Institute in Munich.

From the point of view of my work, the Munich-
Princeton period was the most fruitful. At that time
I did at least three projects that are quoted even today: I finished the one on the black hole, then I did
another on the gravitational lens, and a third on col-
lisions at high energies. It was a good year. I finally
met Oppenheimer. He was a personality who gave
you satisfaction, so to speak. Usually great personalities seen up close are a bit disappointing. Not he:
he gave the impression of living cheek by jowl with
History. The slightest conversational opportunity
was transformed by him: he was a man of extremely
rapid mental processes, with a vast culture, espe-

cially in physics. In my opinion, his creativity was not on a par with these other monstrous abilities of his: Fermi was infinitely more creative, and so was Wigner, who nevertheless was much less brilliant than Oppenheimer.

Levi: Was Oppenheimer born in America?

Regge: Yes. His father owned a small factory, his mother was a painter and had rather good taste. When Oppy was still a child she had gone to Paris to buy some paintings. These paintings were still in Oppy's house when I got there: three Van Goghs, eight Renoirs, and various other Impressionists. Actually I saw only one painting by Van Gogh. Oppy had gotten rid of the other two because he did not consider them good enough. . . . But the one that was left was stupendous, a field with a sky in the background, which was sold before Oppy's death to avoid inheritance problems.

His conversation was fascinating. As I said, talking with him was talking with History; he could never say anything banal. For example: it is well known that he had quarreled with Teller, and when he was given the Fermi Prize the journalists provocatively asked him whether he considered himself a friend of Teller's. He answered, "I did not think of him as a friend before, I do not think of him as an enemy now." His manner of speaking was the verbal analog of Calder's mobiles: a sentence from which hang other balanced sentences, and so on. With him there was always this verbal tension.

Fundamentally he was a sad man. He had realized that he was not creative at the highest level; he was brilliant chiefly at transfiguring the suggestions and ideas of others. Among the other monstrous intelligences people marveled at in those years there was von Neumann, who died in 1957 before my arrival in America. Of him it was said that he performed integrations by heart. All of them had a certain snobbism. They felt they were prima donnas, they loved to *épater* their interlocutors. The Oppenheimer style has been transmitted to an entire generation of physicists: his manner of lighting his pipe, of talking, walking, his mania about French, due to which all of a sudden everybody began to speak French, and the cult of wine. In short, in all these respects Oppenheimer somewhat resembled Philo Vance: long discussions about certain qualities of cognac.

Levi: He was also an expert in Oriental religions, wasn't he?

Regge: He knew Sanskrit, read it fluently. When I moved from the university to the Institute in Princeton I had the opportunity to become acquainted with Oppenheimer and Gödel, whose conversation, on the other hand, was practically nil. He was an extremely reticent man, he did not even go to the Institute's restaurant in order not to see the "crowd," who in fact were only his colleagues and not at all intrusive. Very shy. Once I met him at a dinner, and I think I'm one of the few people who spent a few hours at table with him. I managed to extract something from him; not very flattering comments on Ber-

trand Russell, a few more benevolent opinions on
Peano. I asked him whether he had been part of the
Vienna Circle: he answered with a dry and conclu-
sive "no." He was a close friend of Morgenstern, the
economist, who one day went to see him but found
the house deserted. A pot was boiling on the stove
but of Gödel not a trace. Knowing what he was like,
Morgenstern began to inspect the house and found
him in the cellar hidden behind some sacks of coal,
his phobia of visitors was so great.

Another prima donna was André Weil, the
brother of Simone, a true mathematical genius. Simone used to say that his youth had been comparable only to that of Pascal. He was the type who de-
molishes all his teachers: a supreme mathematician
with a certain horrific character. He lived next to me
and he looked at me with condescension, though at
bottom he thought me pleasant and would consent to
a few brief conversations. I remember the first meeting at the Institute, a very long table with all those
stars, the old professors: a furious verbal brawl exploded between Oppenheimer and Weil because
Oppenheimer had been able to obtain photocopies
of a letter Weil had addressed to the Board of Trust-
ees and of the reply, both very critical, verging on
insult. I would have liked to disappear under the
table. The best part of it is that Oppenheimer had
told me how marvelous the faculty meetings were:
"You've never seen anything like it," he said, and
he was right. At the end he asked me if I had ap-
preciated the spectacle.

I also got to know Heisenberg very well, less well Pauli and Dirac. In Germany I had already had several conversations with Heisenberg: he was a man of enormous culture, obviously very creative. He had carried out a great many studies, not only on quantum mechanics but also on the first theory of magnetized matter and on the "S-matrix," which is very important. Certain original points of view on elementary particles are owed to him. He too ended up at the Institute, with his unified field theory, which had no sequel. His was an evolution parallel to that of Einstein: at a certain point he got stuck in the formalism of his theory and never came out of it. We may say that he ran into his natural limits, as did Einstein and other scientists. There are others, of course, who did not encounter these limits because they did not push ahead. I'm saying that this is a risk run only by those who dare much.

Also Heisenberg did his most important work in his youth, at twenty-six to be precise, when he defined for the first time some very important concepts on measurement in quantum mechanics, which in synthesis is that if I measure an atom, my measuring apparatus disturbs the atom and the result I arrive at depends on the procedure I have applied. In other words, there does not exist an objective reality, or at least not in the traditional sense.

Levi: Was that a completely new concept? Had nobody ever thought of it before?

Regge: Strange though it may seem, only philosophers of little value whom Einstein scorned had dealt with it.

Einstein never accepted the probabilistic interpretation in quantum mechanics. This is a very strange and exemplary story. If one goes back over the developments in quantum mechanics one finds Planck, who calculated the spectral distribution of the black body. If one takes an ideal, perfectly absorbent cavity, and heats it, a gas of photons is created inside it. This radiation has a certain spectral distribution, which is the same as that of a red hot body, a bell-like curve which is Planck's curve. Much was already known about this curve, but it had not yet been calculated theoretically. Planck considered the electromagnetic field as an ensemble of oscillators in which the energy must be proportional to the frequency of the oscillator. . . .

This meant that in light everything behaves as though inside the cavity the energy were distributed in small parcels proportionate to the frequency. This was a result that for a long time remained isolated. Many years later Einstein explained the photoelectric effect by again using light quantums.

This took place in 1905, when Einstein was already quite well known as a physicist. He had published two or three studies of a certain importance on statistical mechanics, and when he published the latter, the general opinion was that he had blundered. But Millikan began to check Einstein's predictions, and after years of hard work his experiments absolutely confirmed them, point by point. Millikan had to concede that despite "the absolute unreasonableness of Einstein's hypothesis," the results proved him right. We have not yet reached quantum mechanics, but Einstein had already understood that there was something wrong in the fact that the energy of light should be concentrated in small parcels, which he did not call photons (the name photon was invented by someone else, in other circumstances, and has entered physics by a roundabout route). He mulled a lot over this concept and in the end he became convinced that light had a particle, which was the photon, and he started a solitary and desperate battle that lasted eighteen years, until 1923. He had to confront people of high caliber, such as Bohr, until the discovery of the Compton effect, from which one sees that light transmits not only energy but also a quantity of motion and therefore behaves in every way like a billiard ball. This brought all opposing theories to their knees.

Levi: The pressure of radiation was not yet known?

Regge: It was known since Maxwell's time that light carries momentum and energy in a particular manner. For some reason, nobody thought about the momentum. The photon, whose discovery is indisputably owed to Einstein, as also his biographer Abraham Pais ascertained in *Subtle Is the Lord*, acted as a destabilizing element in physics. Einstein always had an acute sense of uneasiness due to the duality between field and matter, that is, field and particle. The entities that carry energy and momentum seem to be two: first the particle and then the field, because also the electromagnetic field seemed to carry and absorb energy. The fact that the electromagnetic field had a quantum of light, that is, the photon, in a certain sense removed the duality, demonstrating that the field evinces itself through the photon. With the discovery of the photon, another duality disappears, that between light and matter: light *is* matter, a particulate matter with a certain type of interaction and a certain velocity. But also light possessed a field, and so there has arisen the logical necessity of giving a field also to the other forms of matter that did not have it before. One also asked oneself: if photons come from a field, electrons or protons also ought to possess a field. Part of the subsequent development of physics was to give a field to them, too. Now the word "field" and the word "matter" are equivalent for physicists, if one speaks of intrinsic properties.

The theory of fields is very difficult, but there is a way to make it understandable. You are the police in a great city, with a certain number of men who must follow a certain number of thieves. There are only two ways of doing this. The first is that of saying: "I've identified the thieves; there are only a few. It is sufficient for a policeman to tail the thief, take note of where he is, and inform headquarters, which will decide what is to be done." This is the particle method of describing matter: I take a particle and follow it in its movements. But this is a method that becomes inconvenient when there are too many particles or when the number of particles varies.

The second method, which is that of the theory of

fields, is the method of the British police: it stations
a policeman at every road crossing and he takes
note of all the thieves that go by. In the end, headquarters is able to reconstruct the thieves' movements by elaborating the information, though with a
certain approximation. With the theory of fields one
puts oneself in every portion of spacetime, sees all
the particles that pass through it, and writes different equations that link it to what is happening. His-
torically, it is a bit as if for the photon we had em-
ployed the second method from the beginning,
whereas for other forms of matter we used only the
first method. Modern physics has been completed in
two senses: it has described particles as fields, but
it has also described the photon as a particle. And
by so doing it has removed all asymmetry between
the two descriptions, which are now considered perfectly equivalent.

There is something else. One says, "Fine, the
photon's energy is proportional to its frequency; but
in order to know the frequency I must know the
wavelength, that is, have a wave train sufficiently
long for me to speak of frequency." And this is the
point where quantum mechanics connects up, and
Einstein realized where it led. He didn't like it; he
circled around it, firing broadsides of paradoxes. In
my opinion, if he'd wanted to he could have created
quantum mechanics, but he hadn't the least inten-
tion of doing so. And yet, at a certain point, he realized that there were people who said interesting
things. He read a thesis by Louis de Broglie, in
which it was actually claimed that the quantity of

motion depends on the length of the wave. He found it remarkable, sent a copy of it to Erwin Schrö- dinger, arousing his lively interest. After two or three years of attempts, Schrödinger came up with the equation that bears his name, perhaps the most quoted study that exists in physics. Schrödinger later wrote that Einstein had put him on the right path "with a few observations that reached into in- finity." And thus was attained a first vision of quan- tum mechanics, with this equation that seemed to foresee with astonishing precision the characteris- tics of atoms, to the point of seeming miraculous. It explained the radiation emitted by atoms, its fre- quency, polarization, very detailed structures of the hydrogen atom. Then later on it turned out that this same equation, with a few modifications, accounted for the properties of the atoms of helium, lithium, and several other atoms. In line of principle, all of chemistry is contained in Schrödinger's equation with which it is possible now to calculate a priori the properties of the most complex chemical com- pounds, even of organic compounds: alcohol, amino acids, cellular membranes. From this equation it suddenly became clear that the electron is accom- panied by a wave whose frequency imparts energy to the electron. But there is a wave. What is this wave? We all use it but nobody actually knows what it is.

Levi: How did the research on unification develop?

Regge: At first the unification took place between electro- magnetic forces and weak forces, that is, those that are responsible for the decay of the neutron, in a completely different direction. Einstein was already no longer interested in what a proton, a neutron, all the phenomenology of elementary particles might be. Instead he had attempted the attemptable, to unify gravitation and the electromagnetic field. A real and proper revolution. His discoveries have produced the advances in the field of atomic mechanics, have given origin to very many applications that have refined our technical means, our comprehension. Modern physics is the fruit of this revolution, even though Einstein did not directly participate in this phase. In the end he was completely estranged from this type of thematic. He was approximately fifty when this took place, and yet he still wrote a certain number of very interesting papers on cosmology. But also on unification opinions are mixed. For example, two researchers, Klein and Kaluza, set out on a path of their own toward the unification of forces and in the thirties they sent Einstein a study that he kept in his drawer for two years because it did not interest him. Now that work has been completely reevaluated and is considered much more important. It is the so-called study of dimensional reduction, and postulates the existence of a fifth dimension besides the four of spacetime. An object moves in three spatial dimensions, but should also be able to move in the fourth, even if we do not see it. This fourth dimension must have something very different from the others, if we are prevented from perceiving it in the traditional sense. The reason this perception is hindered is that this dimension is very brief. It is the same as setting out on a path that loops in upon itself, so that one is immediately back at the point of departure. One might compare space to a cylinder, in which normal dimensions run along the cylinder's axis, while the fifth dimension completes a turn around the cylinder itself. An incredibly brief turn, 10^{-33} centimeters, the one hundred billionth part of a nucleus, a length that is evanescent even when compared to the nucleus. To speak of movement in that direction is somewhat strange—one cannot image a microscopic object turning in that direction. We do not have a flying object that allows us to move into this fifth dimension.

What were the virtues of this unification? Particles have attributes, everything has an attribute, physics speaks in a manner worthy of logical analysis: there is a single subject, which is matter, and matter has various attributes, as Aristotle said, position, energy, mass, electrical charge, momentum, and so on. The state of matter is determined by the list of all the various attributes that one can give it. The task of physics is to identify what the possible attributes of matter are, and to establish laws that make it possible to foresee the successive positions. If I have an electron inside an accelerator, and I know the field of the accelerator, it is the same thing, because I can give the beginning of the orbit and be in a position to extrapolate the orbit's subsequent phases. An elementary particle is a fragment of matter so small that it has as attributes only position, speed, and orientation (or spin, like that of a mini-top). It should not therefore exhibit an internal structure. Until a few years ago it was thought that the proton was elementary, but it is not; it is made of quarks.

In particles the most important attributes are position, momentum, energy, charge. Klein and Kaluza present the charge as the component of momentum along the fifth dimension, that is, the charge represents the movement along the fifth dimension. The gravitational field along the fourth dimension appears as the electromagnetic field. With this they were able to establish a formal analogy that is very beautiful. Accord to them, the existence of an electromagnetic field tells us that we live in a space that has five dimensions, in which the gravitational field appears to our primitive senses as an electromagnetic field. The study was then published in the thirties, but nobody noticed it. Now it has suddenly come up again in the context of the new physics. In recent years this technique, which is actually called dimensional reduction, has surfaced again and is being applied on a much larger scale. It is thought that space has eleven dimensions, one temporal and ten spatial. With the Klein-Kaluza technique, as many as seven are peeled away: seven short dimensions that compose a heptasphere, a very complicated object. The description of this heptasphere is a cause for great delight to afficionados, who subject it to interminable calculations.

Levi: Why do we see only three of these eleven dimensions?

Regge: One reason is the one I mentioned. The other is that in order to see these so very brief extra dimensions one must have a microscope so powerful as to resolve so evanescent a fraction of an atom. To make such a microscope one must use very short wavelengths, or, according to Einstein's study, very high energies, having a factor billions of times greater than that of accelerators built up until now. We are completely out of scale. . . . With the present accelerators we resolve details of the subatomic structure that go to 10^{-18} centimeters. We have still to conquer the 10^{-15} factor. It will be an awfully long time before we can directly verify the Klein and Kaluza hypothesis.

　　　　　With the Big Bang, nature may have performed for us the experiment that we do not know how to carry out. At the beginning the universe was much denser than it is now. At a fraction of a second from the Big Bang the density was such that the entire mass of the universe was contained within the dimensions of one atom. The dimensions of the universe were perhaps comparable to the extra dimensions, and there was no longer any difference between the three known spatial dimensions and the other seven. For one ephemeral instant the universe had eleven dimensions. If we had been in there taking measurements we would not have been able to distinguish the usual dimensions from the others.

The concept of number and dimension is anthropomorphic—it is linked to the present epoch, the machines we have, our limited knowledge. If our knowledge were more extensive, we would be able to see other dimensions with very strange properties, provided such dimensions exist, for if they were the same as ordinary dimensions we would see them: we would ourselves be made in four dimensions, and have a perception of it.

Levi: But there is something that doesn't work in this business. If there are other dimensions, they are heterogeneous in respect to the three we know. Considering that physicists love symmetry, this is asymmetry.

Regge: True, and in fact physicists don't much like this point of view. The problem is to understand why certain dimensions remain the ones we know, and others instead have deviated, becoming so short, curving in that way. It is a hotly debated subject that already has a name: spontaneous symmetry breaking. One may hope to issue from it with the field equations: space has an inner tension that renders it unstable along certain dimensions and not along certain others. For the moment, no convincing explanation has been arrived at. Grafted on to this is another thematic. Einstein's great contribution was to say that physics is geometry. General relativity represents the gravitational field as a curvature of space. It is incorrect to say that physics is only geometry. It is geometry plus a principle of action.

It is possible to conceive of infinite different geometries, but what leads us to choose one geometry rather than another? Only a principle of action—in fact, that of Lagrange. We can imagine very many trajectories of a stone's fall, but we select one because we have the equation of motion. The trajectory of motion is what renders a certain action minimal. A ray of light traverses a refracting medium. Why does it follow that path? Why does the optical path become minimal along the course of the ray of light? And let us remember that we ask ourselves similar questions about all particles.

A space extended equally in all eleven dimensions is an unstable space. In a space of that kind a small gravitational field would be sufficient to create a local collapse: this space would very rapidly be deformed toward one of the spaces in which seven dimensions are short and three are long. The study of these instabilities is one of the most interesting chapters of today's theoretical physics. The idea is to demonstrate that only a space made in that manner is truly stable and the other spaces are unstable, inasmuch as they develop within themselves monstrosities that make them die.

However I should not be surprised if in reality space had infinite dimensions, of which we would always see only a finite number. And I should not be surprised if the history that takes place now turned out not to be the most interesting. Perhaps during the first 10^{-40} seconds there may have lived a race of beings with ten dimensions whose scale of energy was immensely higher than that of today, and therefore an inversely much briefer scale of times. Perhaps their existence has been compressed in what for us is an evanescent fraction of a second, but for them it had psychologically the duration of about one hundred million years.

Freeman Dyson has published a study in which he tries to extrapolate what will happen to the world in an extremely remote future. A Dyson born among those hypothetical beings I mentioned before perhaps tried to anticipate the existence of man in a future in which the universe had expanded along a certain number of dimensions. Dyson presents a hypothesis in which matter is stable, in which, that is, the proton does not decay; but if the proton were to disintegrate, matter would disappear and only light would remain. The first immediate phenomenon is the death of the stars: in a hundred billion years, or even sooner, the sun will be ended, will have consumed its fuel; the stars will begin to gather around the center of the galaxy, because the latter is gravitationally unstable. There exists a tendency of the stars to evaporate, so that 10 percent of the stars will leave the galaxy. The spiraling arms of the galaxy will end by disappearing; the nucleus of the galaxy will become ever denser, a gigantic black hole being formed at its center. All around there will be very many black holes, due to the collapse of various stars. All the stars catalyze to the point of becoming iron. If I take this dish, Dyson says, and I leave it alone for an immense time, say 10^{30} years,

it will first of all assume a spheric shape, because
the reciprocal force of gravity of the various parts of
the object induces slow atomic transitions, due to
which an atom moves ever closer to the object's center of mass. Over times that long, any object becomes practically liquid and takes on a spherical
form: strictly speaking, solid objects do not exist.
Over an even longer time, there is always a very
small but finite probability of catalysis of thermo-
nuclear reactions, by which contiguous but not su-
perimposed nuclei of different atoms can fuse and
produce iron with the release of energy. A body of
this kind always develops a little energy, slowly
converting itself into iron. Thus the universe is fill-
ing up with iron balls: the Moon becomes an iron
billiard ball, and so do Earth and the planets. These
billiard balls will end by colliding with one another.
But there exists a process of an even more exasperating slowness that not even Dyson is able to write
as a normal number (if I wanted to write it in a normal way, the entire universe would not be enough).
In a time approximately equal to $10^{(10^{76})}$ years, every
celestial body catalyzes into a black hole and emits
energy as a consequence of this catalysis. In the end
everything is transformed into radiation.

How could humanity possibly survive under such
circumstances? Dyson's idea is that human intelligence would have to migrate into other structures
that use less energy, but that to achieve this aim
they would have to slow down their psychological
perception of time. It would be necessary to arrive

at beings for whom the passage of one hundred years would psychologically mean only the passing of a fraction of a second—enormously extended beings who capture also the tenuous light of very distant stars, and are warmed by it, accumulating energy and then having occasional exchanges of signals internally. A structure of this kind could survive for an indefinite time.

Levi: It's the situation of Hoyle's *Black Cloud*.

Regge: Yes, with the difference that Hoyle's "black cloud" had very fast reaction times, that is, human ones, whereas these beings would be pachydermic, of an exasperating slowness: superdinosaurs. In a science fiction novel written by a physicist and published recently, an expedition of scientists goes to visit a pulsar, that is, an extremely collapsed and dense neutron star. In order to understand its characteristics the scientists hurl X-ray flashes, and they do not know that on the surface of their neutron star there lives a population of beings in a monstrous gravity situation, 10^{13} times the normal one. They are nuclear beings made of molecules on a nuclear scale. If you take a habitat in which energy exchanges are a million times higher than the chemical exchanges, reaction times are also a million times more rapid, and vice versa.

Levi: But it isn't quite clear how the transition from humanity as it is now to that of a rapid type takes place. They wouldn't be our descendants but a completely different race.

Regge: Dyson does not elaborate on this. But one idea would be that of constructing organisms based on superconductors.

Levi: Do you mean beings constructed by us, or able to develop on their own by natural evolution?

Regge: In my opinion it is much more likely that we'll end up by constructing them during the next hundred years. But to get back to the novel, the idea is that of a population that evolves at a rhythm one million times higher than the human rhythm. For them the X-rays emitted by the scientists are normal light, and at very long intervals—what for them is a hundred years—they see a bolt of light appearing in the sky. From that begins a religion, which slowly becomes science. Within a few generations these beings are able to understand the origin of the bolts of light and they themselves will build a spaceship that utilizes a small black hole, they take off, and make contact with the terrestrials. They instantaneously absorb all the terrestrials' technology, and within a few minutes they evolve to the highest level—what we would be able to attain in a million years—and they begin to speak an incomprehensible language. I don't know how the novel ends. Certainly badly. What I wanted to say is that the universe can accumulate in its scale of evolution different structures in drastically different times, each of which may think it is the single aim for the existence of the universe. We think of the Big Bang as of a fraction of time, but in this fraction the most incredible things may have happened: there may have flourished and disappeared the most evolved civilizations, which looked at us as at a distant future. And vice versa. Reading Dyson, I have the impression that what for him will be the distant future will appear to those who live in it as completely normal: we will be the strange and ephemeral beings.

Dyson also elaborated a whole series of hypotheses known as "Dyson's civilization": what can a future human population do if not disassemble the sun or build a cap all around the sun that will receive the sun's light so that it can be exploited completely? At that point humanity will achieve complete control over the sun's light, so that the sun will no longer appear as such but as an infrared star. To search for interstellar civilization in this situation one must identify infrared stars of a very particular type, with very well organized radio signals, which means a very evolved civilization, capable of controlling all the light from a star.

In connection with this theme there arises the usual question as to whether the universe is finite or infinite. For Dyson it is infinite, open. I think his conception is very interesting, because it agrees with certain "theological" preconceptions of mine, which are the following. Let us take Gödel's work. What is interesting, provocative about it? The fact that he speaks to us about the incompleteness of human logic. Human language is finite, composed of a limited number of signs and combinations. The question is whether these combinations can represent the universe around us. The answer is probably negative. According to Gödel's theory, a formal system considered as such cannot serve even as the basis for a complete representation of arithmetic, or a mathematical theory as simple as the theory of numbers.

And there are theorems that cannot be decided. This theme of the incompleteness of man vis-à-vis the universe has meant for many the defeat of "strict" positivism. On the other hand, a neopositivist conception says: "No, I can push ahead as far as I want but I will never arrive at understanding the entire universe." If the universe is infinite, I will be able to understand portions of it that are even greater but still nil when compared to the whole. Man is powerful in the sense that he is able to push the limit always further, but he will never reach a final unified theory. He will always find theories that are unified better, but the final one does not exist, composed as it is of infinite ingredients.

In Dyson all this becomes the universe. Man will be able to see ever larger portions of it, and in fact the visible area is expanding more and more. And yet man must be ready to confront the unexpected: the universe is infinite and he must resign himself to living in a corner of it. That's why I like Dyson's theory. Instead, after a while, with a finite universe we are in a position to see it all, and then the drive toward research ceases, and boredom takes over.

My passion for Borges and the *Library of Babel*, which I reread now and then, is well known. For me the universe is really Borges' library, in which instead of books there are atoms in all their combinations and chemical compounds, and the structure of matter. All the possible ways in which matter can aggregate are dictated by the natural laws, which we still don't know entirely, but it is believed that the theory of fields tells us of forces existing among particles and how they can aggregate. It is like discovering all the possible books that can be written, setting out from a certain number of signs. If the universe is finite, it means that it cannot contain all the possible variations foreseen by theory. I, however, would like the universe to contain or eventually contain in its history, which is infinite, every possible object that can be conceived of by the field equation. A statue of Primo Levi made of Tibetan olive oil cooled to less than two hundred degrees certainly must exist in some part of the universe: it is an extremely improbable object, and it will be necessary to travel a certain number of billions of years in order to find it, but it exists somewhere. The universe is infinite because it must allow for everything that is permitted, because everything that is permitted is obligatory.

What does Borges say? He says that any book that contains a certain permutation is contained in the library. I have calculated the whole thing: the library would be immense, a volume of 10^{3000} cubic kilometers, and so the entire universe would not be enough to contain it.

Levi: It would be huge, but finite. Also in a universe that contains all possible things, it remains to be demonstrated that it must be infinite.

Regge: Probably the number of infinite combinations is infinite.

Levi: One must also decide whether distances are quantified or continuous. If they are continuous, there is no solution.

Regge: One can always find an object made in a certain fashion, provided one moves sufficiently far away. I can think of the strangest object, but I must know that when it is larger than a couple of atoms, and is a somewhat complex molecule, in order to find it I must travel increasingly incommensurable distances. The greater the object, the greater the number of light years that I must travel. For me making a natural law without there being an object that satisfies that law is senseless. If something is foreseen by the theory of fields, that something must exist somewhere. For it to exist it is necessary that the universe have infinite extension and duration. You will tell me that I'm talking religion, but these hypotheses satisfy my esthetic sense. That is why I feel so strongly about Borges' *Library*. Another book by Borges that interests me is *The Garden of Forking Paths*, because it anticipates certain visions of quantum mechanics. Everett and Wheeler's hypothesis appeared about twenty years ago, but it was not successful. The idea is this: Let us suppose we have a particle that disintegrates into two particles. One of these particles can issue on the right, the other on the left, or, with a certain different probability, the two can exchange roles. Only after disintegration will I know where the particles went. According to Everett, what happens is that the two possibilities exist. It is as though the history of the universe were bifurcating into two parallel universes, so that both can exist. When we try to control the existence of something, since our observatory is contained in the universe, we have the impression that all that happens occurred in the branch in which we live. Whoever is in the other ramification, naturally, has the impression that things went as he saw and measured them.

Both probabilities each have the impression of being the supreme and indisputable truth. In reality, if we were able to go outside the universe, we would discover that it has become bifurcated, that there are two or more parallel histories. This is a bit what happened in Borges' story, which in fact is cited by Everett and Wheeler. It is a metaphysical answer, let's say, because it cannot be tested.

Levi: A possible idea, that cannot be demonstrated in any way.

Regge: Metaphysical, precisely. It cannot produce experiments, because it places itself beyond physics and imparts to the matter an exclusively intellectual disposition. For a certain number of years it has been said that the scientist must not concern himself with such things: theology in its broader meaning is not respectable, philosophy is an extracurricular ornament. Dyson restored its value; he always took heterodox positions and he distinguishes most attentively between a a scientist's professional work and his vision. In a certain sense what had given strength to modern physics is precisely having been able to distinguish between the two moments. One must, after all, have something that drives one, motivates one, induces one to try to demonstrate that the world is made in a certain manner. Naturally I must then set about verifying whether my hypothesis is true or not. If I do not carry out this verification with extreme attentiveness from a scientific point of view, my result is worth nothing. I must follow the rules, and impose a certain discipline on myself.

To those who reproach scientists for being too cold, I always answer that a formula is not cold and is not all there is to science. That would be like reducing a doctor's work to the blood analyses, which really supply the data for an interpretation that goes far beyond. Also, in physics the calculation represents only the point of departure from which one starts in the attempt to go beyond.

Levi: The fact remains that at this point it seems completely senseless to me that anyone who is not a physicist should write a science fiction novel. By now science fiction literature is a private hunting preserve, something written by physicists and for physicists. The part that enters the commercial market is marginal dross. True science fiction is what circulates in the republic of physicists. It was founded by people who knew some physics and biology, which they were able to communicate, and in fact it was very successful.

Regge: But it has aged frightfully. If you reread the "Urania" novels it seems impossible that in twenty years science fiction should have been surpassed by the facts to such an extent. All the trips to the Moon have become ridiculous, they look like papier-mâché scenery.

Levi: Except for Wells's book, which I think is entitled *The First Men on the Moon* and has a very beautiful idea: there is the usual slightly mad scientist who invents a plastic substance that intercepts gravitational pull. If you are on top of a slab of his "cavorite" you no longer have any weight. The scientist decides to build a very simple vehicle to go to the moon, a polyhedron five or six meters in diameter made of small "cavorite" blinds that shelter the inventor and his partner. The vehicle obviously has no weight, and in order to take off a window is opened toward the moon, that is, it "weighs" only toward the moon.

Regge: Wells's astuteness was that he did not try to explain the machines he built. They were far beyond the technological possibilities of his time and therefore they could not age.

Levi: Not always: in *War in the Air* Wells actually foresaw the Second World War, the alliance between Japan and Germany, the preeminence of aerial warfare. By comparison, Verne appears a bit pathetic, a bit didactic. But a novel of Verne's that is not science fiction, *Michael Strogoff*, and which I recently reread, seemed very beautiful to me. And the same should be said about *Around the World in Eighty Days*, which with its forgetting of the time zones is very witty. The fact remains that a popular genre such as science fiction is imploding, and becoming a sort of private hunting preserve, the monopoly of scientists.

Regge: Asimov has stopped writing, but even his famous trilogy contains a fiction that doesn't stand up: a journey at a speed greater than that of light, and since I'm a relativist this sort of thing gives me psychic traumas.

Levi: The demands of the story forced him to do that. One either stops writing or tries to smuggle in elements that go against known laws.

Regge: Light is desperately slow, for science fiction authors: they try to circumvent this fact without ever being able to find a solution. And I suffer.

Levi: Arthur C. Clarke tried to prepare a list of things that are imaginable and of others that are not: the intelligence of mammals, conveyor belt roads, air cushion vehicles all function. Others do not, like the time machine.

Regge: I can understand a journey to the closest star, not in the short term, let us say in three hundred years, provided that humanity enters an extraordinary period of economic development, is able to colonize the band of asteroids, and develops nuclear fusion satisfactorily. The means could be the Orion project imagined by Dyson: a spaceship that has a hemispherical copper mirror inside, enormous, something like ten kilometers, and at regular intervals launched propulsive units, that is, hydrogen bombs, that strike the sphere's center, explode, and propel the spaceship by the wave of the impact. Two hundred thousand hydrogen bombs would be enough to arrive at 1/100 of the speed of light. In this way in four hundred years one would arrive at the nearest star. The fact is that a colony of people aboard an object of this kind degenerates rapidly, regresses to a provincial culture, then encounters severe genetic problems, palace revolutions, and so on. Our knowledge about the workings of a human society are much more imperfect than what we know about machines. Without mentioning the fact that it is impossible to guarantee that the ships' machinery necessary for survival will be able to function for thousands of years. Nowadays an elevator doesn't even have a two months' guarantee.

My idea is more modest. Not a direct journey, which is of no use. Circumsolar space is filled with the nuclei of cold comets, huge spheres of frozen methane and ammonia. It would be sufficient to disseminate throughout the universe colonies living on enormous cylinders, and that when they meet these comets use them by means of nuclear fusion, extracting energy from them to build another colony and propel it a little farther. The comets are like refueling stations. So from nucleus to nucleus it would be possible in the end to arrive at the nearest star. There is not much difference between the nucleus of a comet and the asteroids we know. The asteroid has the same material imprisoned at shorter distances from the Sun, from which certain more volatile elements have evaporated. So much so that Chiron, between Saturn and Uranus, probably is the nucleus of a comet.

These space stations inside the cylinders could house up to twenty thousand persons, and if the distance between one and the other were not excessive there might be exchanges between colonies, somewhat like what happened among the various villages of the Far West. Of course it is necessary to arrange matters so that one cylinder is able to build others, as in the society of the bees, where one colony gives life to other colonies. At the beginning survival cannot be guaranteed, and the cylinders may meet up with failures or suffer damages or malfunctions due to internal or external circumstances. It's a risk. What is there in space? The future of humanity might be there, in the sense that free from gravity, in immense spaces, with a practically inexhaustible quantity of material, if one were in a position to control nuclear fusion one could tranquilly eat up all of Jupiter's satellites. Not Saturn's rings, which are an extraordinary tourist attraction. . . . Saturn's rings contain as much water as one thousandth of all the terrestrial oceans. They are unique structures composed of thousands of concentric rings; in certain instances three intertwined rings have been discovered. The tourism of the future will take place there.

Levi: I've accepted everything connected with space exploration, but I'm surprised that it is possible to survive reasonably well in the absence of gravity, something completely unexpected. The discomforts seem to consist in a bit of seasickness, loss of calcium, things that are ridiculous when compared to the leap from gravity one to gravity zero, or almost. A gift, a present. It is easier to move in the absence of gravity than to reach a depth of ten meters under water.

Regge: Did you ever hear about Gerry O'Neill? He wants to put colonies at the so-called Lagrange point of the Earth-Moon system, that is, at the point of the lunar orbit, sixty degrees out of phase with the Moon itself, where theoretically if an object is left there it remains stably linked to the movement of Moon and Earth. A sort of gravitational cradle. It is thought that this is the best spot to set up a station because it would not be disturbed again and it would not be necessary to adjust its position, as happens with any object placed in any orbit whatever. It is thought that at the Lagrange point there is dust accumulated over millions of years. There are people who claim they have glimpsed its reflection at night. . . .

For protective panels, the cylinder station would utilize material collected on the Moon and shot off by a centrifuge or by a catapult. It is sufficient to impart a speed of one kilometer per second to any object for it to be able to leave the Moon. Then it only has to be oriented correctly, captured in the most appropriate manner when it arrives at destination, and recycled for the defense of the satellite. In this way the Moon will be taken apart piece by piece even though disassembling all of it will take billions of years. Certainly, there would be conspicuous gravitational effects. But it seems that the Moon is already now moving away from the Earth, and therefore the tides will diminish in any case. Also, the Earth is slowing down; billions of years from now a day will last 40 hours, or perhaps more. There has already been one effect: two years ago the clocks were put back by a tenth of a second over the span of a year. The conquest of the Lagrange point would be the first step toward the colonization of the asteroids. An asteroid constituted of an alloy of iron and nickel would be most attractive from a commercial point of view. It would not be a very dense crystal: having condensed in the absence of gravity, in the void, it would be easy to detach in blocks and send to Earth.

Levi: One would have to go quite a distance to find these commercially viable asteroids made of iron. The closest are made of the same material as the minor planets—silicates, basalts. The moonstone is one of our stones. The asteroids between Mars and Jupiter are not well known yet, but I think that they too must contain current minerals. But I would like to return to what Gödel said. To those who reproached him for having destroyed the solidity of mathematics' foundations, he answered that, on the contrary, he had reinstated the value of the role of intuition. Speaking of intuition, what did all that is not scientific culture—that is, art, literature, and music—give you besides Borges?

Regge: I have an anomalous relationship with music. I am a maniac for classical music. I ignore rock, which actually gives me a pain physically. Perhaps I love music so much because it wasn't taught in *liceo*. But my father had me take piano lessons and something has remained with me. I play the piano badly, but I enjoy myself. For me music has to do with some kind of unconscious elements. Physicists, but not only they, prefer Bach. I don't feel up to explaining or quantifying this preference: what the trade of physicist may have to do with Bach's music I really don't know. Perhaps a certain natural rhythm, or perhaps, as has been repeated for years, the Bachian construction is mathematics. Perhaps deep down Bach had intuitions of a mathematical type, that however cannot be adequately represented by the laws of harmony. The same in a certain way can be said about Mozart. I cannot do without music, but I would not know how to link it explicitly to my trade. What is certain is that if I could reincarnate myself I would choose to be a musician.

As for literature, I set out on the wrong foot, as I already said. At any rate, I was very interested in

Mann's *Doctor Faustus*, which in fact centers on the theme of music. And also *The Man Without Qualities* by Musil, who had had a mathematical education. Another mathematician is Lewis Carroll, the author of *Alice*. In physics there is a concept that is called the "Carroll group," and it describes a world at the limits, fictional, in an attempt to imagine what would happen if one were to look at a phenomenon on a scale in which the speed of light is nil. It's been called the "Carroll group" in homage to the famous paradoxical image of the rabbit who is always running and remains in the same place.

Levi: As far as my experience goes, I must say that my chemistry, which actually was a "low" chemistry, almost culinary, first of all supplied me with a vast assortment of metaphors. I find myself richer than other writers because for me words like "bright," "dark," "heavy," "light," and "blue" have a more extensive and more concrete gamut of meanings. For me "blue" is not only the blue of the sky. I have five or six blues at my disposal. . . . I mean to say that I have had in my hands materials that are not of current use, with properties outside the ordinary, that have served to amplify my language precisely in a technical sense. Thus I have at my disposal an inventory of raw materials, of tesserae for writing, somewhat larger than that possessed by someone who does not have a technical background. Moreover, I've developed the habit of writing compactly, avoiding the superfluous. Precision and concision, which, so I'm told, are my way of writing, have come to me from my trade as a chemist. And so has the habit of objectivity, of not letting myself be easily deceived by appearances. This is the story of the potassium that I told in *The Periodic Table*: sodium very much resembles potassium, but when you get close to it you realize that there is a difference.

But these are only personal impressions. The question that is often addressed to me by my readers in high school ("If you hadn't been in a Lager and hadn't studied chemistry, would you still have written? And if so, in the same way?") could be given a sensible answer only by taking another Primo Levi, who didn't study chemistry and set out to write. The convalidating proof does not exist. Sometimes, slightly straining the paradox, I've written that my model of a writing style was the short end-of-week report, and to a certain extent this is true. I was struck by a sentence attributed to Fermi, who also found it boring to write compositions in *liceo*. The only composition he would have written gladly would have been: describe a two-lira coin. Something like that happens to me: when I have to describe a two-lira coin, I'm quite successful. If I must describe something indefinite, for example a human personality, I'm less successful.

Also, the habit of weighing words, of not trusting approximate words—these are the rules of cooking, nothing abstract: before using a word one must investigate its scope and its linguistic area. Then one can draw from the greater reservoir of artisan languages, which allow one to rediscover an ancient linguistic mechanism. The major part of rhetorical figures that have by now entered into the common language comes from departmental languages, from the language of the mill, the stable. Now there is a great number of other linguistic sources, such as factories, banks, travel. Why not use them? In this sort of operation my previous trade serves me well: for me, an ex-chemist, "to filter" says something more than it says to the layman. There are many concepts that have a precise meaning and can be conveniently exploited, with effects that are perhaps different from those I expect but have an effect nevertheless. Everyone says "to crystallize," using the word improperly, because the original concept has been lost, at least in part.

Another virtue that the chemist's trade develops is patience, not to be in a hurry. Today chemistry is completely changed; it is rapid chemistry. Today the analysis of a mineral is no longer manual. It is done by machine and takes a few minutes, where before it took weeks. Naturally, it was inconvenient having to work a whole week to analyze the mineral, but this made it possible to develop other virtues, which in fact are those of perseverance, not getting discouraged, assiduous application. Now it happens that just as in the elementary schools a discussion is going on as to whether children should be taught how to use a pocket calculator instead of doing their arithmetic manually, for the chemist the question arises whether it is worthwhile to teach the manual analysis method, so-called systematic analysis, that requires a lot of time and also a lot of material. Now a tenth of a milligram is enough. It's put in the machine, somewhat like what happens with clinical analysis, which is capable of coming up with an electrophoresis in the space of a few minutes. In the past, making a separation of proteins was very hard work. It is clear that from a practical point of view a machine-made analysis is more convenient. But manual analysis, like all manual work, has a formative value; it is too similar to our origins as mammals to be neglected. We should after all know how to use our hands, our eyes, our nose. I'm very glad that I educated my nose: I'm able to identify by smell certain functional groups more quickly than the infrared spectrometer and the gas chromatographer. In this way the physicist's trade integrates you in your function as a complete person who does not neglect any of his potential faculties. Whether as a student or a varnish technician, I made a continuous effort to exploit fully the help that eyes, fingers, and nose can afford.

With all this, after describing how much I owe to my old trade, and my gratitude for what it has given me (including a pension and the opportunity to travel on business to half of Europe), I must add that the day I left was for me a day of liberation: I thought I was walking on clouds, as you said about passing your maturity exam. A purer, less contaminated liberation than the one that put an end to the Lager, and had taken place against a backdrop of

slaughter and death. I spent the day after my resig-
nation strolling through the streets of Turin on a
working day: a working day—do you realize what
that meant? No more office hours, no more crossing
town during the rush hours; and every blessed day,
no night calls because a valve has broken or a rain-
storm has flooded the cable beds. I felt I had ava-
lanches of free time at my disposal: if before I had
written three or four books, working in the evening
and on Sunday, now I would write another twenty or
thirty. Instead it didn't go like that: a friend of mine
used to say that in order to do things, "one mustn't
have time." Time is an eminently compressible material.

Certainly, I wrote again, afterwards, also drawing
on my factory experience as in *The Monkey's
Wrench*; but now I have the impression that I have
exhausted the reservoir. I write poetry, without
much believing in it; as for prose, it seems to me
the time has come to go up a new path, both in
terms of theme and language. But for the moment
I've gone back to working on a collection of essays
on the Lager "revisited" after forty years, that I had
started a few years ago and interrupted to write *If
Not Now, When?* But this undertaking has crossed
with an unforeseen fact: I read Pozzoli's book *Writing With a Computer*, and it had on me the effect of
the bugle call that wakes you up in the barracks. I
realized that today one can certainly live without a
computer, but one lives at the margins and is bound
to become more and more detached from active so-

ciety. The Greeks said of a person without culture: "He can neither write nor swim." Today one should add: "Nor use a computer."

So, without hesitation or undue inner effort, I bought myself a computer, and now I write exclusively with it. At the beginning it was tough: I was totally ignorant of the terminology in use, I was terrified by the fear that what I had written would in the end be erased because of some mistaken move, and the instructions in the manuals seemed to me indecipherable. Then, little by little, I understood certain fundamental things. In the first place, that one must repress the human desire to understand what's inside: haven't we used the telephone for almost a century now, and the television for forty years, without knowing how they work? And do we perhaps know how our kidneys and our liver work, which we've been using since always? It's purely a matter of habituation; for the rest, I'm told that, except for the specialists, not even physicists and mathematicians bother to delve deeper. They've domesticated the marvelous monster and use it without anxieties. In the second place, I understand that it is senseless to hope that one can use the device by studying the manuals. Apart from the fact that their language is irksome (those gnomes have invented portentous things but did not take care to create alongside it an agile and expressive language), it would be the same as expecting to learn how to swim by reading a manual without ever going into the water—indeed, as my son the biophysicist says, without even knowing what water is, without ever having seen it. One must learn in the field, making mistakes and then correcting oneself.

I'm still a neophyte: I still have to learn a lot of maneuvers, but it would already cost me a great effort to go back to the typewriter, or, worse, ballpoint, scissors, and glue. I do not exclude that the new instrument exercises a subtle influence on style. Long ago, having to engrave letters one by one with hammer and scalpel forced us to be concise, to adopt the "lapidary" style. The labor has been reduced step by step, and now it is almost abolished: you compile, correct, polish, cut, insert parts in a text with ridiculous ease. In short, we have reached the opposite extreme. It seems to me that this facility tends to encourage prolixity. I'll have to be careful.

There is also another danger: my equipment does not only write, it draws. I haven't drawn anything since my fourth year in elementary school, and I get an almost indecent amusement out of creating on the small screen forms that seem to me beautiful and new, much above my manual ability; and then I can "put them in memory" and print them. It is so fascinating that I waste many hours at it. Instead of writing, I play, and enjoy myself like a child. And besides, I'm glad to learn how to do these new things at sixty-five.

Regge: Questions about the electronic future and the computer revolution are among those I hear most often.

I own a very small PC, the smallest and least expensive. I use it for research only marginally. I find it hard not to be captivated and drugged. It's the perfect toy for adults, much better than the electric train, officially given as a present to the children but in practice the monopoly of the parents.

In my opinion the computer will end by spreading like a prairie fire and will change the face of the world, at least the industrialized world. It is impossible to resist. And if we should arrive at a machine that is truly intelligent or simulates intelligence reasonably well, then we'll really have to look out.

To those who ask me for advice on how much computer time they should allow their children, I can only advise to avoid automated and preprogrammed games, which mortify the intelligence and make us slaves of the machine. I look with favor, as a true intellectual gym, on the programming and consequent control of the machine. The calculator amplifies human possibilities; it amplifies intelligence but of course also imbecility.

Are calculators really intelligent, at least those used by scientific institutions? I'm unable to consider them that. I and also my colleagues are used to programming them, controlling them in the slightest details, telling them meticulously what they must do. For me it is still a machine—docile, meticulous, and tireless, completely devoid of imagination and self-determination.

Perhaps the next generation of computers will present itself to us with automated characteristics but with the consequent loss of control over the machine. In such a case we will have the illusion of intelligence. I don't even exclude that I am a biological computer who thinks he has a personality precisely because my inner logical circuit continues to repeat it to me every time I ask it. I'm not always convinced that I present a continued illusion of intelligence.

Levi: What you said before about the eleven or infinite dimensions of the universe, the forked or multi-forked universe, the colonizing of the comets, the disassembling of the Moon, have had the effect of an overdose on me. I was about to interrupt you and ask you where does science end and science fiction begin; but then I understood that the limit no longer exists, or rather, it is indefinite and shifts from year to year. You are the true masters of the world. What will become of the human species during the coming years depends on you. The power you have is boundless. You understand more things than the layman, and all of energy is in your hands, in both its aspects, the beneficent and inexhaustible (it should become almost free of charge, right?) energy of nuclear fusion and that of the heads that ride on intelligent missiles. What do you think of the conclaves physicists hold every year in Erice?

Regge: When we speak about Erice it is necessary to distinguish between the talks that have been held there— not always wise and not always felicitous—and the idea per se of bringing together scientists of a bellicose type, gathered from all parts of the world. I am very much in favor of the idea, and I think that Zichichi has sensed just in time the necessity to continue the dialogue among the great powers outside the by now worn-out channels of classical diplomacy.

I was perplexed by the declared intentions of certain American extremists and war-mongering elements, in which reference is made to the construction of a very extensive antimissile satellite, whose cost would amount to a trillion dollars. This superweapon, which has been compared to the one that appears in *Star Wars*, is supposed to utilize a laser effect but in X-rays, the whole thing fueled by the explosion of an atomic warhead. As a physicist, I was surprised by the very hint at the possibility of optic pumping in the region of X-rays. I believed it was reserved for a possible but distant future. Instead, it seems to be within reach. Regrettably this is a discovery that introduces a further element of instability and wild expenditure in a world that certainly has no need for either.